SOLVING WORD PROBLEMS

NOW YOU CAN!

BY JENNIFER HARRIS

ISBN: 978-1-943258-96-3

Editing: Amy Ashby

Published by Warren Publishing
Charlotte, NC
www.warrenpublishing.net
Printed in the United States

This book was created to help parents, teachers, and students master the process of solving word problems. Let me show you how.

CONTENTS

"The only way to learn mathematics
is to do mathematics."

–Paul Halmos

1

READ THE PROBLEM

I READ CAREFULLY AND NOW I UNDERSTAND THE TOPIC!

- Read the problem quickly the first time. Then, read the problem slowly the second time. It is also helpful to read the problem out loud when possible.

- Read the problem as many times as desired until you understand the topic.

- You will know you understand the topic if you can explain it to someone else.

- Make sure you underline or highlight key words.

"If you define the problem correctly,
you almost have the solution."

–Steve Jobs

2
WHAT DO YOU SEE?

SALLY HAS *eight* APPLES
AND BILL HAS *two* APPLES.
HOW MANY DO THEY HAVE
all together?

- Identify all given information. That means pick out the most important parts.

- A word problem will always give you something to work with. It may give you numbers, variables, an equation, a formula, or other important math facts.

- Make notes about key facts.

- Break the word problem up into puzzle pieces or parts. Print out the word problem and then cut words or phrases into pieces.

- Draw a picture to help you see what you have been given. Throw out the parts that are not important.

"Math is like going to the gym for your brain. It sharpens your mind."

–Danica McKellar

3
DETERMINE WHAT YOU'RE BEING ASKED TO DO

HMMM ... WHAT DO I NEED TO DO NEXT?

- If you read the very last sentence of the problem, you will be asked to give an answer to something.

- Make sure you understand what you are finding the answer to. Sometimes a picture is helpful, so sketch a drawing if possible.

- Determine what your answer should look like.

- Is your answer going to be a specific type of number? Is your answer going to be in dollar amounts? Miles? Feet? Seconds? Number of people?

- Are there multiple steps needed? If so, make a list of the steps.

"It's okay not to know. It's not okay not to try."

–Author Unknown

4
ARE THERE THINGS YOU DON'T UNDERSTAND?

- Are there words you don't understand? If so, define what these words mean and how they help with the word problem.

- Are there things you are supposed to remember how to do in order to solve this word problem?

- If so, what are they? If need be, ask a teacher or parent, or search online for how to do them.

- You absolutely must understand every word in the word problem.

"Good mathematics is not about how many answers you know … it's about how you behave when you don't know."

–Anonymous

5
ARE THERE MATH FACTS OR FORMULAS NEEDED?

AREA = LENGTH X WIDTH

EX: 4 ft X 12 ft

AREA = 48 sq. ft.

DISTANCE = RATE X TIME

PERIMETER = 2L + 2W

A (B+C) = AB + BC

HERE'S AN EXAMPLE:

3 (2+10) = 3(2) + 3(10)

= 6 + 30

= 36

- Are there any math facts or rules needed? If so, what are they?

- Look at the page to the left. These may be familiar formulas to you. Dependent upon the problem, you may need additional formulas as well.

- What formulas do you need to solve the word problem?

- Sometimes multiple formulas are needed.

- If you don't remember the formula, hop online and type the name of the formula in the search bar.

- Multiple websites will show the formula and give advice on how to use it.

Never say, "I can't."
Always say, "I'll try."

6
WHAT MATH MUST BE PERFORMED?

%
x 4
1
5+9
6
Y
7-4

- What operations need to be performed?
- Will you need to add, subtract, multiply, divide, use exponents, etc. Is there a specific set of steps that should be followed?
- Are you going to perform multiple math operations? If so, what operation are you going to perform first? Second? And so on?
- Be clear as to which items will receive your mathmatical "operation." Add what? Subtract what? Multiply what? Divide what?

“The person who does the work is the *ONLY* one who learns.”

–Harry Wong

7
PERFORM THE CALCULATIONS

- Use the information from previous steps and math formulas or math operations to solve the problem.
- Go ahead. Do the math.
- I have confidence in you.
- You can do it!

Mistakes are proof that you are trying.

8
REVIEW YOUR WORK

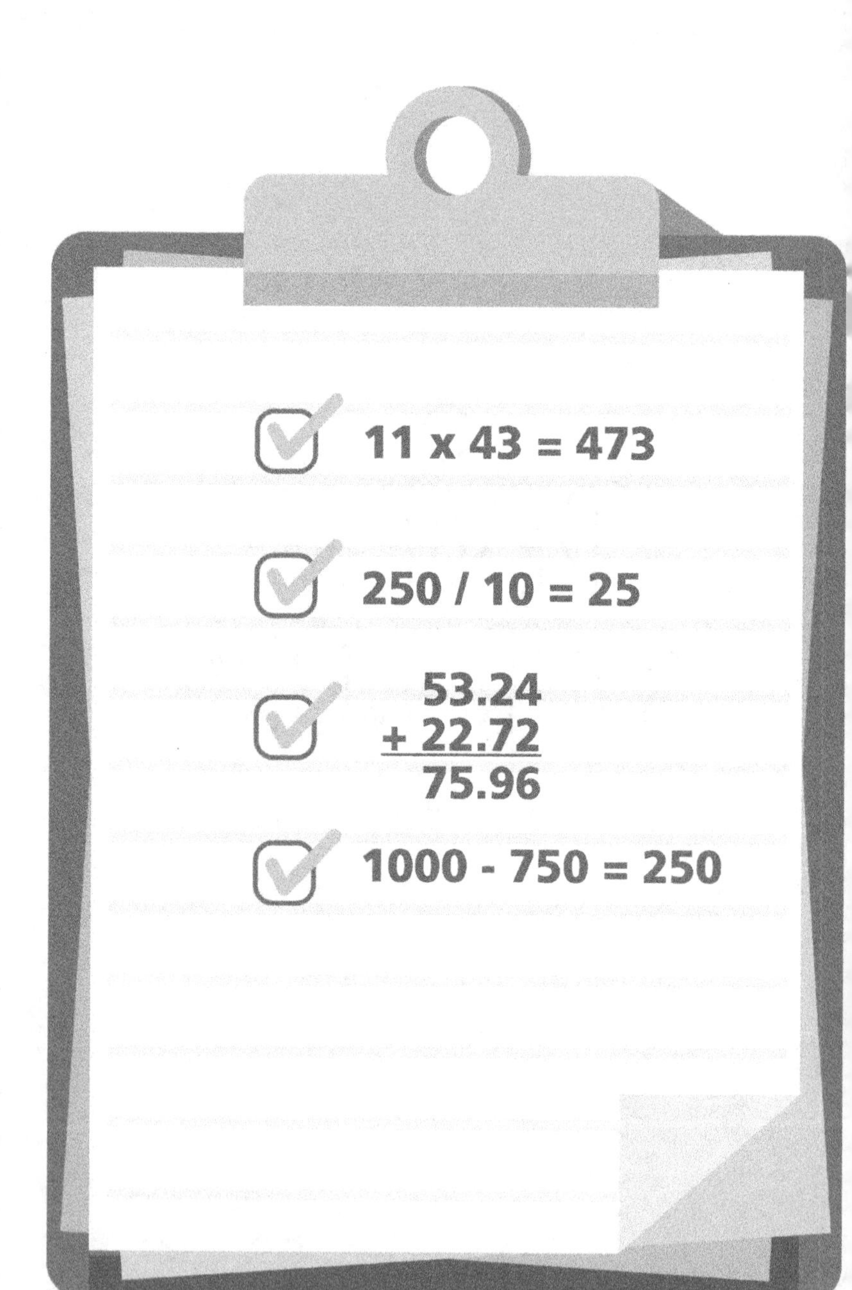
11 x 43 = 473
250 / 10 = 25
53.24
+ 22.72
75.96
1000 - 750 = 250

- Review or check your work.
- It's easy to make careless mistakes, so checking your work is always a great idea.
- Sometimes you can plug your answer back into the problem to check your answer.
- Did you perform the correct operation?
- Did you plug in the proper numbers?
- And finally, did you answer the question?

"It is clear that the chief end of mathematical study must be to make the students think."

–John Wesley Young

9
MAKING SENSE OF IT ALL

- Does your answer make sense?
- Can you make a complete sentence with your answer to the question?

    Example 1: Answer = 10
    Sally and Bill have 10 apples all together.

    Example 2: Answer = $100
    She earned $100 at work this week.

Don't ever be afraid to ask for help.

10 EXPLAIN THE PROBLEM

- Can you explain or demonstrate the problem to another person?
- Talk to someone else about the word problem.
- Explain to them what the word problem is about.
- What did it ask you to find, and how did you decide to solve this problem?
- After you explain or demonstrate the problem, ask your friend if they have any questions about this word problem.

"If you stop at general math, then
you will only make general money."

–Snoop Dogg

11
TAKE IT A STEP FURTHER

- Create your own math word problem that is similar to the one you just completed.
- Make it into a story that includes you as the main character.
- This will help you connect with the word problem and deepen your understanding of math in general.

Your receipt
Effective 7/17/2020 - Per statewide orders, library users are required to wear masks when inside library buildings. The exception to this rule is for children 10 and younger. If you forgot yours or do not have one, we can provide one for you while supplies last.

Items that you checked out

Title: Solving word problems : i just don't get it--now you can! / by Jennifer Harris.
Due: Saturday, February 20, 2021

Total items: 1
Account balance: $0.00
1/30/2021 4:06 PM
Ready for pickup: 0

Questions? Call 1-888-861-READ(7323) or visit us at www.MyLibrary.us

TIME TO PRACTICE

1 Read	**Sara has 10 apples and Paul has 13 apples. They share 8 apples with Carl. How many apples are left?**
2 What do you see?	10 apples 13 apples > and Share 8 apples
3 What are you to find?	How many apples are left?
4 Any questions?	What does "share" mean? It means to take from someone and give to another person.
5 What math facts or formulas are needed?	Remember: How to add numbers. How to subtract numbers.
6 Math to be peromed	Add Sara and Paul's apples, then subtract 8 apples to share with Carl.

10 + 13 = 23 Then 23 - 8 15	**7** Do the math.
Did you use the numbers from the word problem? Did you add then subtract?	**8** Review
There are 15 apples left after Sara and Paul share with Carl.	**9** Make a sentence.
Explain it to another person.	**10**
Create a similar problem.	**11**

1 Read	**Jackie is 2 years younger than Bill. The sum of their ages is 68. How old is Jackie?**
2 What do you see?	We don't know Jackie's age but we know it is 2 years younger than Bill. We don't know Bill's age. When you don't know the value of a number, you choose a variable (a variable can be any letter of the alphabet) to temporarily represent it. So we chose: x = Bill's age $x - 2$ = Jackie's age Sum of their ages = 68
3 What are we to find?	We need to find Jackie's age.
4 Any questions?	What does "sum of their ages" mean? It means to **add** their ages. What does "2 years younger" mean? It means to subtract by 2.

Remember: How to add like terms. How to solve for X. (Search the Internet if you forgot.)	**5** What math facts or formulas are needed?
Set up an equation to solve using previous steps. Bills age + Jackie's Age = 68	**6** Math to be performed
$x + x - 2 = 68$ $2x - 2 = 68$ $+2 = +2$ $2x = 70$ $2x / 2 = 70 / 2$ $x = 35$ (Bill's age) $35 - 2 = 33$ (Jackie's age)	**7** Do the math.
Bill's age + Jackie's age = 68 35 + 33 = 68 68 = 68	**8** Review
Jackie's age is 33.	**9** Make a sentence.
Explain it to another person.	**10**
Create a similar problem.	**11**

1 Read	**Find the area of the sector of a circle with a radius of 100 miles, formed by a central angle $\pi/2$ radians.**
2 What do you see?	Area of a circle. Sector of a circle. Radius = 100 miles Central angle = $\pi/2$ radians Now draw a picture of what you know so far. 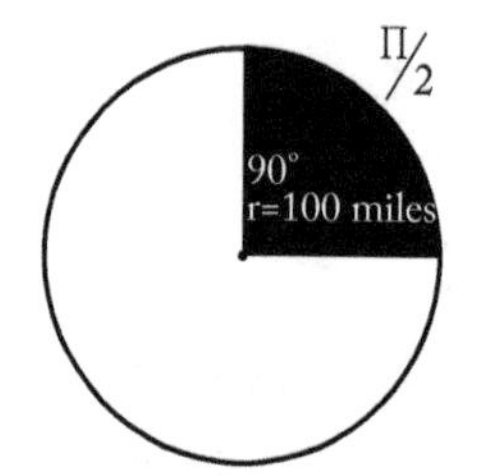
3 What are you to find?	Find area of the shaded sector of the circle.
4 Any questions?	How do you convert radians to degrees? How many degrees are in a whole circle? How do you create the ratio for a portion of the circle?
5 What math facts or formulas are needed?	Use $A=\pi r^2$ Remember: $\pi/2$ radians = 90° (central angle)

Convert radians to degrees.

$\frac{\Pi}{2} \times \frac{180^\circ}{\Pi} = 90^\circ$

Central angle converted to degrees.

The sector (90°) is what ratio of the circle (360°)? Create the ratio.

Use ratio above and $A = \Pi r^2$ to calculate area of the sector.

6
Math to be performed

$\frac{90^\circ}{360^\circ} = \frac{1}{4}$

Remember: Area of a circle. We don't need the area of the whole circle, just ¼ of the circle. (i.e. the shaded section.)

$$A \text{ shaded section} = \frac{90^\circ}{360^\circ} \Pi r^2$$
$$= \frac{1}{4} \Pi \cdot 100^2$$
$$= \frac{1}{4} \Pi \cdot 10000$$
$$= 7853.982 \text{ miles}^2$$

7
Do the math.

Review work.

8
Review

The area of the sector is 7853.982.

9
Make a sentence with your answer.

Explain it to another person.

10

Create a similar problem.

11

ABOUT THE AUTHOR

Jennifer Harris is a mathematics lecturer at Clayton State University in Morrow, Georgia. She has more than fifteen years of experience teaching math to students ranging from sixth grade to college. Her goal is and has always been to help students master mathematical skills and their application to real life by making math simple and easy to understand.

If you are interested in having Jennifer teach a workshop at your school or organization, you may contact her at:

jenniferharris2@clayton.edu

RESOURCES

www.funeasymath.com
www.maththatmakesense.com
www.maththatmakescents.com